THE CAVENDISH SQUARE ANIMAL LIBRARY

The World of FISH

BY JUDY THORPE

Cavendish Square

Published in 2026 by Cavendish Square Publishing, LLC
2544 Clinton Street Buffalo, NY 14224

Website: cavendishsq.com

Cataloging-in-Publication Data

Names: Thorpe, Judy.
Title: The world of fish / Judy Thorpe.
Description: Buffalo, NY : Cavendish Square Publishing, 2026. | Series: The Cavendish Square animal library | Includes glossary and index.
Identifiers: ISBN 9781502675125 (pbk.) | ISBN 9781502675132 (library bound) | ISBN 9781502675149 (ebook)
Subjects: LCSH: Fishes–Juvenile literature.
Classification: LCC QL617.2 T46 2026 | DDC 597.156–dc23

Editor: Caitlin McAneney
Designer: Deanna Lepovich

The photographs in this book are used by permission and through the courtesy of: Cover (image) Damsea/Shutterstock.com; cover and various pages (wave vector) Olga_TG/Shutterstock.com; various pages (tape element) Tamara Luiza/Shutterstock.com; p. 4 Nick Utchin/Shutterstock.com; p. 5 VectorMine/Shutterstock.com; p. 6 DR. Zafer Haddad/Shutterstock.com; p. 8 Dudarev Mikhail/Shutterstock.com; p. 10 Kazakova Maryia/Shutterstock.com; p. 11 I Wayan Sumatika/Shutterstock.com; p. 13 (top) MIRAI Think OK/Shutterstock.com; p. 13 (bottom) Drew McArthur/Shutterstock.com; p. 14 Martin Prochazkacz/Shutterstock.com; p. 15 Gena Melendrez/Shutterstock.com; p. 16 Goinyk Production/Shutterstock.com; p. 18 MoniqueLivesLife/Shutterstock.com; p. 20 Noheaphotos/Shutterstock.com; p. 21 Kletr/Shutterstock.com; p. 22 Simon Shin kwangsig/Shutterstock.com; p. 23 Maquiladora/Shutterstock.com; p. 24 Jack FotoVerse/Shutterstock.com; p. 25 blue-sea.cz/Shutterstock.com; p. 26 Tatjana Kabanova/Shutterstock.com; p. 28 Joe Belanger/Shutterstock.com; p. 29 UW.ART/Shutterstock.com; p. 30 Somphob Boonlaim/Shutterstock.com; p. 31 Sergey Novikov/Shutterstock.com; p. 32 AZ Septian/Shutterstock.com; p. 33 (top) Sekar B/Shutterstock.com; p. 33 (bottom) Ronnie Chua/Shutterstock.com; p. 34 LisaAngeline123/Shutterstock.com; p. 37 Cynthia A Jackson/Shutterstock.com; p. 38 Arturo Vigil/Shutterstock.com; p. 39 VisionDive/Shutterstock.com; p. 40 LukaSkywalker/Shutterstock.com; p. 41 thediver/Shutterstock.com.

CPSIA compliance information: Batch #CSCSQ26: For further information contact Cavendish Square Publishing LLC at 1-877-980-4450.

Printed in the United States of America

CONTENTS

INTRODUCTION

Fish are incredible animals that live in bodies of water all over the earth. More than 24,000 species, or kinds, of fish exist, from tiny fish smaller than your fingernail to superlong sharks. There are many differences between the fish species on Earth, but they all have a few things in common. They have a backbone, which makes them vertebrates.

This image shows fish of many sizes!

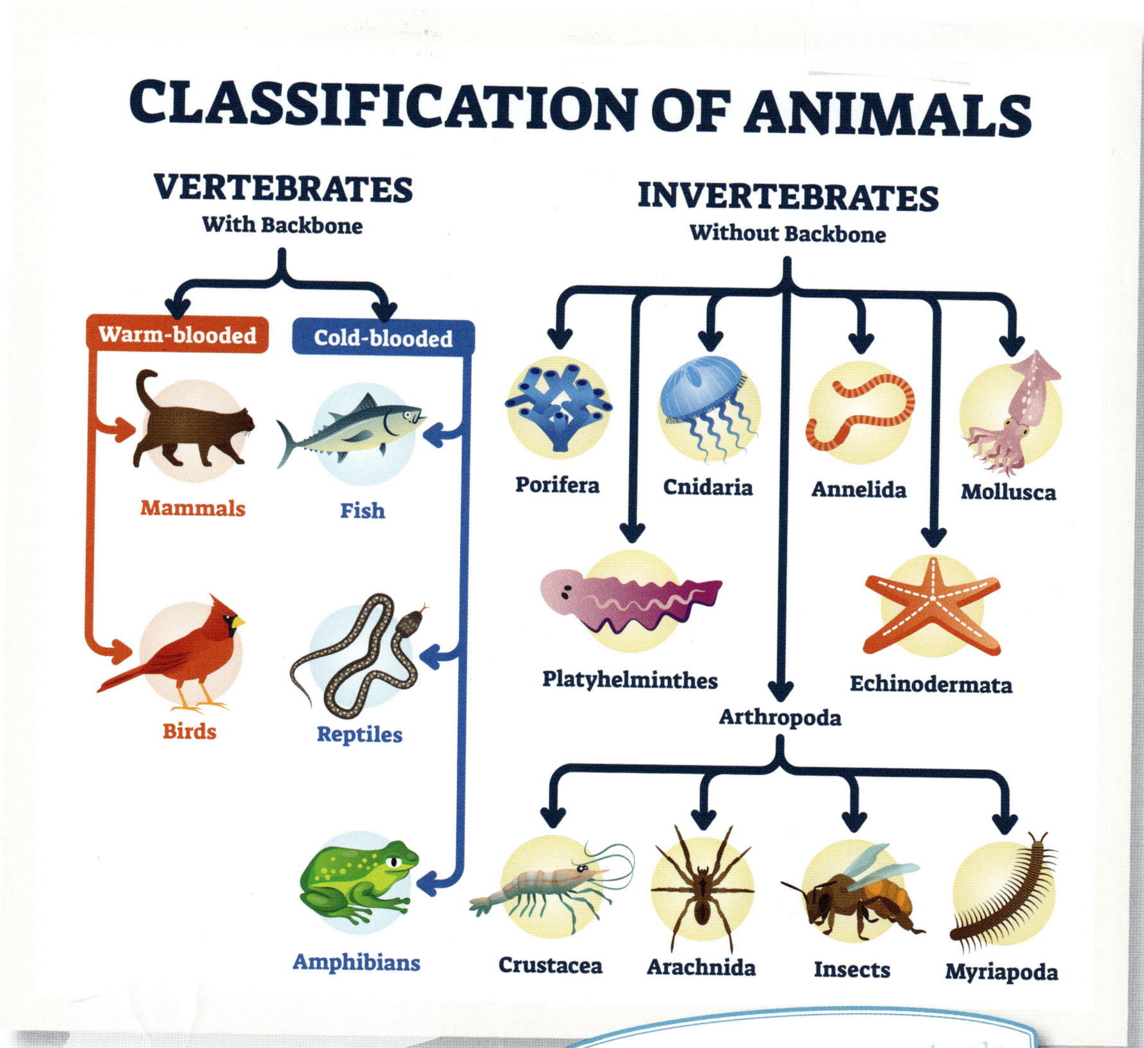

This chart shows how animals are grouped. It includes some scientific names!

They are also cold-blooded, which means they get their body temperature from their surroundings.

Fish have developed great **adaptations** to live in the water. They don't need to breathe oxygen in the air like we do. Instead, they have gills. Gills allow fish to get the oxygen they need from the water. Different species of fish have additional

A seahorse might not look like a typical fish, but it is still a fish!

adaptations, from their body shape to the color of their scales, to deadly defenses that keep them safe from predators.

The world of fish is rich with **diversity**. A seahorse looks nothing like a stingray, and a deep-sea anglerfish looks nothing like a tube-shaped trumpet fish. What a fish looks like and how it behaves are often influenced by its **environment** and the part it plays in its ecosystem. For example, fish that live in tropical ocean waters need different adaptations than fish that live in cold lakes and streams. In addition, fish that are mainly predators often behave differently than fish that are mainly prey for larger aninmals.

There is so much to discover about the underwater world of fish, so let's learn about the incredible fish that fill Earth's waters!

Fish have bodies built for life underwater.

CHAPTER 1

FISH FEATURES

Fish have been around for nearly 480 million years. They were the first vertebrates to live on Earth, **evolving** in shallow ocean waters. Over time, fish developed the fishy features we observe today. You can distinguish fish from reptiles and mammals quite easily. But it can be harder to identify which species of fish you're looking at. You can note their size, shape, surroundings, and colors to give you clues.

As you look at a fish's features, it's good to ask yourself: Why might this feature help that species survive? Does it seem like a predator, prey, or both?

THE FISH KINGDOM

Which features do all members of the fish kingdom have in common? As you learned in the introduction, they have a backbone and are cold-blooded. Most of them also have fins that help them move around in the water. Caudal fins (also called tail fins), on the backside of some fish species, help propel fish in the water. Pectoral fins on the sides of the body help change direction and speed. Dorsal fins on the top of the body help with balance.

Most fish also have scales. These small plates on the skin of a fish help protect them from harm in the water. They also help them move through the water. Some fish have very thick scales that act as armor. Some even have teeth on the edges of their scales that can keep other animals from harming them.

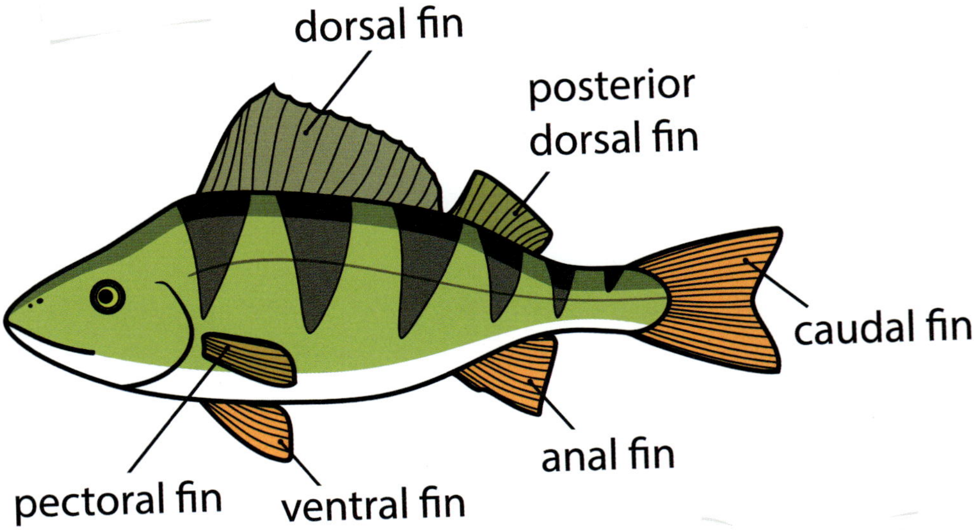

Shown here are different kinds of fins. Ventral fins are found behind the pectoral fins. Anal fins are found closer to caudal fins.

THAT'S WILD!

Catfish have no scales. Most species just have skin, and some can even breathe through their skin!

Oxygen is a gas all animals need to live. It's present in the air. But how can fish get oxygen if they can't breathe air? That's what their gills are for. Oxygen passes from the water through the gills and into the fish's blood. Some fish, including mudskippers and some kinds of "walking" catfish, can breathe

Mudskippers can stay on land for days at a time!

for short periods of time on land through their skin.

DIFFERENT GROUPS

There are three main groups of fish—bony fish, sharks and rays, and jawless fish. Each have their own special features and abilities.

Bony fish have hard skeletons made of bone. Most of the fish in the world are bony fish, from small clown fish to huge blue

> **THAT'S WILD!**
> There are over 400 modern families within the bony fish group, and more than 29,000 species within them!

marlins. These are usually the fish you think about when you think about going fishing. They also have scales and a swim bladder that helps them with **buoyancy**. Bony fish are found in all underwater environments, even the deep sea!

Sharks, skates, and rays have flexible skeletons made of cartilage, which is similar to the material that forms our ears and noses. Rays include stingrays and manta rays. They have flattened bodies shaped like pancakes. They flap their "wings" to "fly" through the water. They're known for their sharp barbs or **venomous** spines on their tails that can hurt or kill other creatures. The giant manta ray can grow to 26 feet (8 m) across!

Sharks have been around for 300 million years. More than 300 species exist today, from the gentle giant, the whale shark, to the fearsome great white shark. They're often fast swimmers, using their torpedo-shaped bodies to cut through the water. They live in different kinds of ocean waters, and some species are at the top of the ocean food chain.

Jawless fish also have skeletons made of cartilage, but instead of jaws they have suckerlike mouths with small teeth. These are some of the oldest species of fish on Earth. Jawless fish include lampreys and hagfish. They attach themselves to other fish with their sucking mouths and eat into the flesh of their prey. Sometimes their prey dies, and sometimes it's just left with a large wound. They are considered parasites because they live off another creature.

BUILT FOR THE WATER

Many of the features seen in fish equip them for life in the water. Most fish have streamlined bodies that help them slip

A stingray has its gills and mouth on the bottom of its body, with its eyes on top. That helps it hide in the sand on the ocean floor.

SHARK SENSES

Sharks, rays, and skates have a "super" sense. With special organs called ampullae of Lorenzini, they can pick up on electrical signals from other creatures. These jelly-filled sense organs can pick up on even the weakest electrical fields. This helps sharks and rays find their way through water, even when it's dark and murky. It also helps them forage, or look for food. Any movement—even a twitch—from a creature can be picked up by animals with this super sense.

In sharks, the ampullae of Lorenzini are pits on their heads. Sharks also have a great sense of smell. In fact, around two-thirds of a shark's brain is just used for smell! Some can smell blood from far away, alerting them to bleeding prey for an easy meal.

Great white sharks are feared all over the ocean for their amazing hunting abilities—thanks to their super senses!

Lampreys have circular rings of pointy teeth.

easily through water. But the shapes of the bodies are very different. A tuna is shaped like a teardrop. Flounders are flat like pancakes. Eels are long and thin like snakes.

The shapes of their bodies sometimes give fish special hunting or hiding abilities. For example, moray eels hide in the nooks and crannies of coral reefs, waiting for smaller fish to swim by. Then the eels dart out and grab their prey with strong jaws lined with needle-sharp teeth.

A fish's features help it move through water with the speed and agility that they need in their specific environment. A typical fish swims by wiggling the rear part of its body and tail from side to side. It curves its body into a S shape with its

It's easy for these flat, colorful reef fish to swim around the corals.

muscles. A fish moves its fins to hold itself steady, to turn, and to stop. To change depth, a bony fish uses its swim bladder. This is a sac filled with air inside its body. To go up, the fish lets more air into the swim bladder. To go down, it lets air out.

Different kinds of water environments require different fish features. For example, deep-sea fish are built for life in the deep, dark, cold waters of the ocean. These fish, such as anglerfish, can create their own light. This is an ability called bioluminescence. On the other hand, tropical fish found in coral reefs have their own special features, such as thin bodies for weaving in and out of corals and bright colors to help them blend in.

The place a fish occupies in its food chain also determines some of its features. For example, fish that are often lower on the food chain, serving as prey for other animals, often have adaptations that allow them to hide. Fish that are near or at the top of their food chain, acting mainly as predators, often have adaptations such as sharp teeth, strong jaws, and venomous body parts. These features make them expert hunters.

Parrotfish are herbivores, or plant-eating fish, using their strong teeth and jaws to scrape algae off coral reefs.

CHAPTER 2

FISH ABILITIES AND BEHAVIORS

It's a fish-eat-fish world under the water. Every day is a test of survival, whether fish are living in a small lake or the ocean. Fish need to use their senses to steer clear of threats, including predators, as well as find enough food to live. Some fish are social, while others are solitary. Some are fierce hunters, while some live off plants.

Fish have evolved over millions of years. Their unique, or special, adaptations, including their abilities and behaviors, are key to their survival in their environment.

FISH SUPER SENSES

Fish need special senses to find their way through the water. Unlike humans, whose eyes are set inside their heads, a fish's eyes bulge outward. This gives them a wide field of vision. A fish can see on both sides of its head, behind, and in front—all at the same time!

A fish has ears, but they are inside the fish's head, behind the eyes. Sounds travel very well through water and easily enter a fish's body.

THAT'S WILD!
A fish has taste buds in its mouth. Some fish also have taste buds on the outside of their heads and fins.

Imagine what you could see if your eyes bulged outside of your skull like a fish!

Fish use sounds to attract mates and to warn off rivals.

A fish has nerves that form a line along each side of its body. This line is called a lateral line. The nerves detect tiny differences in pressure around the fish. Lateral lines help fish swim close together in schools without bumping into one another.

A fish smells using nostrils located on the sides of its head. Some sharks can detect the scent of one drop of blood in an Olympic-size swimming pool, which helps them hunt. Meanwhile, Pacific salmon use "odor memories" to find their way from the ocean to the river where they were born.

Catfish have taste buds on their whisker-like feelers.

FISH SCHOOLS

Larger fish are often solitary, or live alone. They don't need the protection of a larger group—or any help hunting. This is true for most shark species, as well as the large sun fish and deep-sea anglerfish.

> **THAT'S WILD!**
> Schooling fish use their lateral line to sense their neighbors' movements. If a few fish sense an attacker and change patterns, the others will move accordingly.

However, smaller fish may swim in large groups of hundreds or thousands. These groups are called schools. Swimming in a school gives each fish extra protection. It is less likely that one fish will be attacked when it has other fish around it. Also, a school of fish has more eyes to spot predators.

Fish use less energy when they swim in groups. This is because a fish creates water currents when it swims, giving

These small, silvery fish gain protection from predators by swimming in a big group.

a boost to others next to it. Plus, fish in schools can find more to eat and feed on larger kinds of foods than if they were swimming alone. There is strength (and success) in numbers!

FISH THAT HUNT

Fish hunt in many different ways. Some follow and attack prey. Some hide and **ambush** prey. Others use electricity or venom.

Some fish, such as the tuna and great white shark, spend their days swimming through the water looking for prey. They are big fish, and at the top of their food chain. Sharks have rows of razor-sharp teeth to snag prey and rip it to pieces.

Shown here are shark species in order of size. The megalodon, pictured at the top, was the largest fish ever known—but is now **extinct**.

THAT'S WILD!

Bluefin tuna have the sharpest vision of any bony fish on Earth and are very fast. They use these abilities to hunt schools of fish.

Other fish hunters aren't as fierce. Remoras, also known as suckerfish, attach themselves to larger fish and turtles using a sucking disc on the top of their heads. When their hosts eat, the remoras catch the leftover scraps. Little fish called cleaners feed on the dead skin and tiny creatures that live on other fish. These other fish sometimes line up as they wait to be cleaned.

Large fish even let the cleaners enter their mouths and clean their gills.

A few fish have very unusual ways of getting their food. The electric eel of South America kills or stuns its prey just by touching it. The eel's muscles produce an electric current. Anglerfish use "lures" that come off the tops of their heads to interest their prey, then snap them up. That allows them to use as little energy as possible—while scoring a huge meal.

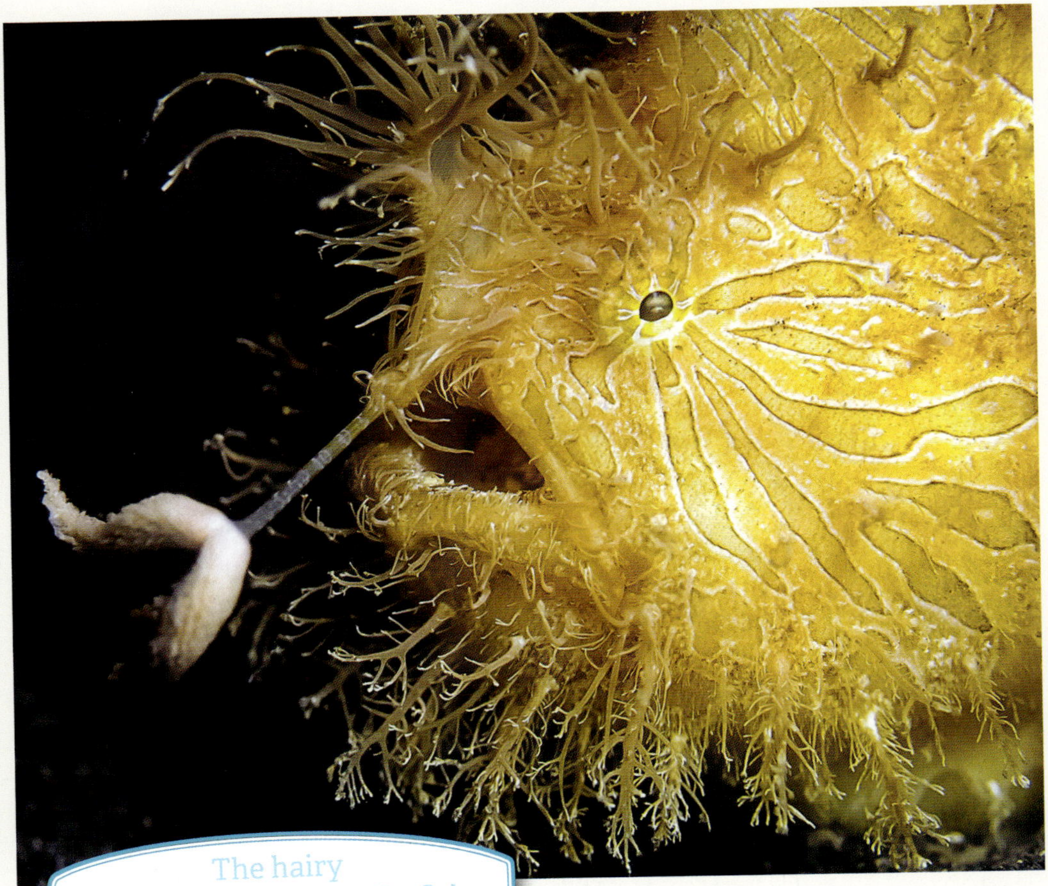

The hairy frogfish, a kind of anglerfish, can snap up its prey in a fraction of a second.

ON THE DEFENSIVE

Some fish have adapted to ocean life by developing amazing defenses. For example, the hagfish—which has survived for 300 million years as a species—oozes slime out of its body, which chokes predators before they can take a bite.

A porcupine fish sucks in water and makes itself three times its normal size when a predator comes near. As the porcupine fish puffs up, spines on its skin stick straight out. This makes the fish too spiky to eat.

Other spiky fish include the stonefish and lionfish—but their spines are actually venomous. In fact, the stonefish is one of the most venomous creatures in the world. It has 13 spines on its back that let out venom when it is under pressure or in defense.

Can you spot the stonefish? Not only are they venomous, they are masters of blending in.

Catshark eggs are released into the ocean in a leathery pouch sometimes called a "mermaid's purse."

CHAPTER 3

A FISH'S LIFE

The way fish grow and develop is as different across species as their many features and behaviors. Some fish are viviparous, which mean that babies develop inside their mothers and then are born live. Some are ovoviviparous, which means eggs develop within the mother, but then the babies come out of the eggs before exiting the mother's body.

Most fish, however, are oviparous, which means that mothers let out eggs that babies have not yet come out of. Mother fish often have no role in raising their babies after leaving their eggs behind. These baby fish are left to go through their life cycles on their own.

IT STARTS WITH AN EGG

THAT'S WILD!
Baby fish are called fry.

Most fish mothers take a fins-off approach to making babies. Female oviparous fish lay hundreds or thousands of eggs in the water. Once in the water, these eggs are **fertilized** by males. The eggs range in size from tiny specks to pea-sized blobs. When the eggs hatch, the newborn fish are usually on their own. Only a few survive to become adults.

See-through fish eggs give us a glimpse into the development of the baby fish inside.

Some fish care for their eggs. Sunfish scoop out nests on pond bottoms and protect their eggs. Male clown fish guard their eggs from predators. Female African cichlids carry their eggs in their mouths. After the babies hatch, mothers keep them in their mouths for several weeks.

Grunion are small fish that live along the southern California coast. On certain spring nights during high tides, they leave the water to lay and fertilize their eggs on sandy beaches. The sand keeps the eggs warm. About two weeks later, the baby grunion hatch and are carried out to sea by the waves at high tide. They'll spend the rest of

THAT'S WILD!

People can predict when the "grunion run" will be each year, and many people flock to the beaches to watch these silvery fish make their way to the water.

Male clown fish find a safe spot near their home **anemone** to protect their eggs.

their lives in the water until they come on land to **spawn**.

LIVE YOUNG

Some species of sharks and rays give birth to live young. The female shark's eggs are fertilized within her body. Some shark babies will develop inside eggs while inside their mothers. This keeps them safer from predators than fish eggs floating out in the ocean. After developing, the baby shark hatches inside its mother, and then it is released into the water as a fully-formed baby, which looks like a smaller version of its mother. Sand tiger pups, or babies, hatch out of eggs inside their mother but then

THE FATHER OF THE SEAS

Seahorse dads protect their eggs like no other fish in the sea. They take unfertilized eggs from the female seahorse into their special belly pouch, where they are fertilized. A pouch can hold up to 2,000 eggs! While they're in the pouch, the eggs develop thanks to the oxygen and food provided by the father.

After between two weeks and four weeks, the baby seahorses are born. They go off on their own. Their father and mother often mate with only one another for the rest of the breeding season. They do a dance with each other, which includes a change of color, as a form of "courtship."

Seahorse males are often called "the Father of the Seas" for their hard work carrying their eggs.

aren't born for a little bit. During that time, they snack on eggs inside their mother—even their unborn siblings!

Whitetip sharks are born live. They look like smaller versions of their parents.

Blue sharks and bull sharks are two species that are viviparous. Just like mammals, these babies and mothers are connected by an umbilical cord in the mother's body that allows a baby to receive the oxygen and food it needs to develop. After giving birth, the mother shark swims away and the tiny sharks are left to their own devices. However, some shark mothers leave their babies in areas with few predators and lots of food, called nurseries.

THAT'S WILD!

A great white shark baby can be 5 feet (1.5 m) long at birth!

A SALMON'S LIFE

Each summer, schools of sockeye salmon head from the Pacific Ocean toward the west coast of Canada. More than two years before, these fish were born in a lake or stream miles from the coast. They made their way to the ocean as small fish. They have

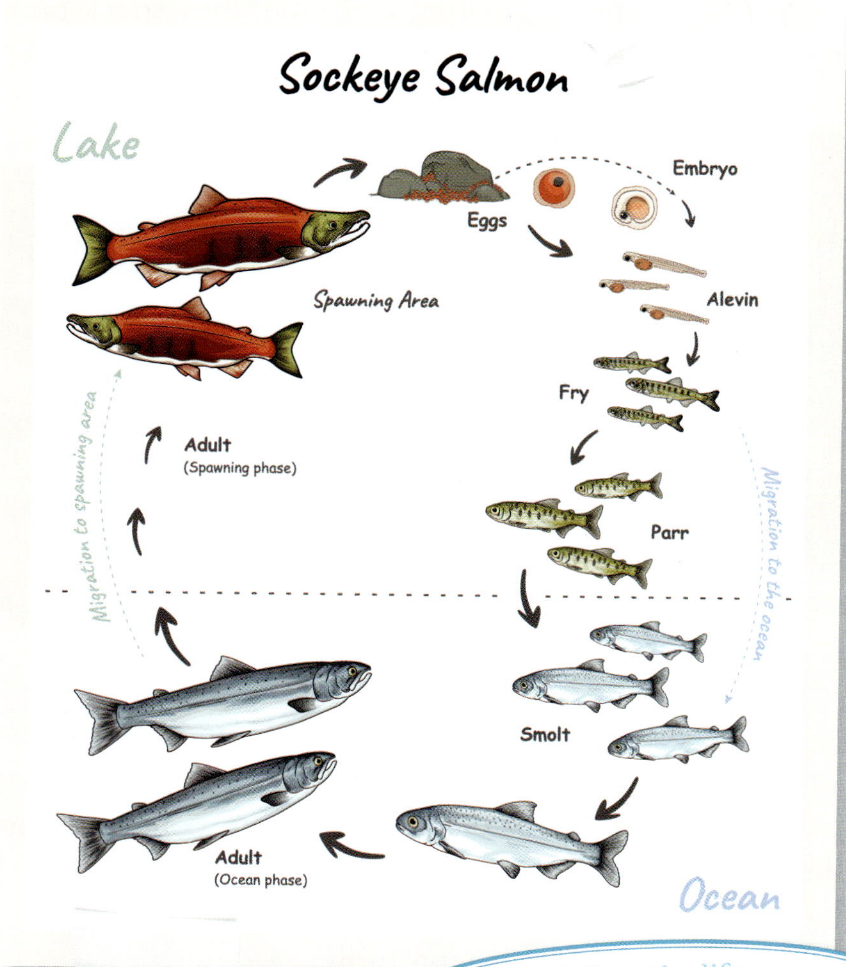

Follow the life cycle of a sockeye salmon with this diagram.

been far out in the Pacific Ocean ever since, finding food to grow larger.

The school swims hard, not even stopping to eat as it reaches the river. The salmon sense their home stream. They know they must swim up the river—against the current—to reach the place where they were born.

Once the salmon leave the ocean, changes take place in their bodies that allow them to switch from living in salt water to living in fresh water. Sockeye salmon change color only after they

Sockeye salmon can do more than just swim upstream—they can swim up waterfalls!

leave the ocean and enter a river. Their silvery, black-spotted sides turn red, while their shiny heads and tails become green.

By fall, the salmon school has traveled up the river. The females dig nests and leave their eggs there for the males to fertilize. The eggs will then develop all winter. In spring, the fry will hatch and stay in the fresh water for one to two years until they're ready to make the trip out to the ocean.

These salmon eggs are nestled between rocks to keep them safe.

Despite their large size, Mekong giant catfish are mostly plant-eaters, living off algae and other small plants at the bottom of the Mekong River.

CHAPTER 4

FISH AT HOME

Fish are built to survive in their own unique environments. Some live in fresh water, while others live in salt water. Some survive in brackish water, which is the slightly salty water in places where rivers and seawater mix. Some, like sockeye salmon, go between fresh water and salt water at different times in their lives.

From fish living in the coldest Arctic waters to those at home in warm coral reefs, these creatures have adapted to their surroundings in some amazing ways. And each species has its unique challenges ahead as Earth's waters warm up and people continue to pollute, build, and overfish.

FRESHWATER HOMES

You may think most fish species live in the vast ocean, but actually almost half of all fish species are found in fresh water, such as lakes and rivers. Around 18,000 freshwater species have been discovered to date.

Freshwater fish such as trout and salmon live in cool lakes and streams. Other fish, such as perch, bass, and pickerel, swim in slower-moving rivers and lakes where water plants hide their food—minnows, frogs, and insects.

> **THAT'S WILD!**
> The largetooth sawfish has been found in salt water, brackish water, and fresh water in warm areas of Earth. It can be 25 feet (7.6 m) long!

Catfish have feelers around their mouths that help them find food in muddy water at the bottom of lakes, rivers, and ponds. Catfish can live in water that is too warm and polluted for most other fish. Some of the largest freshwater fish on record are Mekong giant catfish caught in Thailand, which can be more than 600 pounds (272 kg). Other freshwater giants include sturgeons, giant freshwater stingrays, and largetooth sawfish.

Unfortunately, many freshwater fish are in danger of becoming extinct. They are threatened by pollution and other human activities. Some in the Great Lakes suffer due to **invasive species**, such as the sea lamprey, and overfishing.

REEF FISH

As one of the most diverse habitats in the world, coral reefs are bursting with fish species. These shallow, tropical waters include amazing underwater structures made of the skeletons of coral polyps. They are filled with hiding spaces. Eels hide in crevices and between corals. Strange stinging sea creatures called anemones attach themselves to coral reefs, and their swaying arms act as a home for clown fish.

> **THAT'S WILD!**
> Coral reefs are home to almost 25 percent of all ocean species, even though they make up less than 1 percent of the whole ocean.

Most reef fish are very colorful. They use colors and patterns to attract mates, warn other fish away from their territories, and blend into colorful backgrounds. Their colors make it easier to spot their own kind for mating. They also may confuse predators against the colorful reef backdrop.

AT HOME IN CORAL REEFS

The Great Barrier Reef, off the coast of Australia, is the largest coral reef system on Earth. Over 2,500 reefs make up this system in the Pacific Ocean, and all of them are a little different. More than 1,500 species of fish live here, as well as around 400 species of coral and other sea creatures, including seabirds, sea turtles, and crustaceans. Some parts of the reef are in very shallow water, while others are in deeper waters.

The Great Barrier Reef faces many challenges today, from pollution to climate change. The crown-of-thorns starfish has been a very destructive creature in the reef, eating the living coral. This invasive species threatens the future of some of the reefs in the system. Many scientists are working hard to keep this reef system going for years to come.

The Great Barrier Reef is one of the few living structures that can be seen from outer space!

Some reef fish form schools and feed on tiny living things in the water. Some eat the soft parts of the corals. Some crack open snails and sea urchins. Some larger fish, such as barracudas and sharks, stalk smaller fish. These animals are top predators in the coral reef ecosystem.

Barracudas have thin, torpedo-shaped bodies and sharp teeth. They are perfectly built for hunting in reef environments.

AT HOME IN THE OCEAN

Animals in the open ocean need their own special adaptations to survive. Because there's nowhere to hide, many of these fish rely on camouflage, or blending in. From above, their blue or gray backs blend in with the color of the water. Looking from below, their silver or white undersides blend in with the sunlight.

Some fish of the open ocean, such as the tuna, are fast. They speed through the water, eating smaller fish. But the two largest ocean fish are slow. The whale shark, the biggest fish of all, is as long and heavy as a school bus. It swims very slowly, eating only tiny creatures. It strains them from the water with hairy bristles in its mouth.

This shark has a darker color on top of its body and a lighter color on the bottom. That helps it blend in with the open ocean.

The ocean sunfish, the largest bony fish, can grow to be 10 feet (3 m) long. It swims even more slowly than the whale shark. It spends much of its time at the surface. There it lets seabirds pick tiny creatures off its body.

While sharks and tuna are often found in the top layer of the open ocean, there are many other fish that live deeper. Only the top 1,500 feet (457 m) of the ocean gets any sunlight. It is inky black below this point, which stretches down

10,000 feet to 15,000 feet (3,048 to 4,572 m). These dark depths are home to some of the most unique fish in the world.

This illustration shows the fish found at different depths of the ocean.

Hunting can be a challenge when it's so dark! Deep-sea fish called anglerfish wiggle glowing lures from stalks on their heads to attract prey. Other fish known as gulpers or pelican eels have big jaws and can swallow prey their own size or larger.

Many fish living in this dark world are black with spots that glow in the dark. They feed on shrimp and other fish attracted to their lights. Each kind of glowing fish has its own pattern of lights. These patterns may help fish find each other to form schools. This bioluminescence is a very helpful adaptation. These fish are a shining example of the **resilience** that exists in the wild world of fish!

DEALING WITH CHALLENGES

The fish in Earth's freshwater and saltwater ecosystems all have unique challenges in today's world. One of those challenges is overfishing, or when people take too many of one kind of species. It disrupts the balance in food chains. For example, the giant bluefin tuna can grow to be 12 feet (3.6 m) long, weigh up to 1,500 pounds (680 kg), and live as long as 30 years. They are important predators in the ocean ecosystem. But fishing fleets capture so many bluefins that only a few have time to grow into large fish, which affects the ecosystem around them.

Oceans also suffer from climate change, as the ocean has taken in about 90 percent of the heat from greenhouse gases. The water's temperature is rising, which causes ice melting, sea-level rise, and **acidification**. Countries around the world are working together to fight climate change and protect ocean ecosystems.

Scientists with the World Wildlife Fund believe that the only solution to overfishing will be to make large sections of the world's oceans off limits to commercial fishers.

THINK IT THROUGH

1. Why do you think sharks have survived for around 300 million years?
2. What are the benefits of fish leaving their eggs versus protecting them?
3. Why do you think salmon leave their home waters to live in the ocean?
4. Why is it helpful for fish to be brightly colored in coral reefs?
5. How does the overfishing of one species affect the ocean ecosystem?

GLOSSARY

acidification: A condition when soil or water becomes too acidic.

adaptation: A change in a body that makes it better suited to certain conditions.

ambush: To attack by surprise.

anemone: A group of sea animals without backbones that look like flowers and have stinging tentacles.

buoyancy: The ability to float.

diversity: The state of having many different kinds of things.

environment: The conditions that surround a living thing and affect the way it lives.

evolve: To change or develop over time.

extinct: Died out.

fertilize: To provide male cells to female cells to allow babies to develop.

invasive species: A kind of animal that grows out of control outside of its native habitat, harming other animals there.

resilience: The ability to overcome challenges.

spawn: To produce young, especially in large numbers.

venomous: Able to inject poison into another creature.

FIND OUT MORE

Books

Clasky, Leonard. *20 Things You Didn't Know About Stingrays*. Buffalo, NY: PowerKids Press, 2024.

Emminizer, Theresa. *Fish Life Cycles*. Buffalo, NY: Enslow Publishing, 2025.

Humphrey, Natalie. *Seahorses Are Strange*. Buffalo, NY: Enslow Publishing, 2023.

Mallory, Louis. *Sharks Sense Electricity!* Buffalo, NY: Gareth Stevens Publishing, 2024.

Salvatore, Kaitlyn. *Fish*. Buffalo, NY: Britannica Educational Publishing, 2025.

Websites

Great White Shark

kids.nationalgeographic.com/animals/fish/facts/great-white-shark

Learn more about great white sharks with *National Geographic Kids*.

Monterey Bay Aquarium: Animals A to Z

www.montereybayaquarium.org/animals/animals-a-to-z

Check out all the underwater animals that live at the Monterey Bay Aquarium, many of which are fish.

Pacific Lamprey

www.fs.usda.gov/detail/npnht/learningcenter/kids/?cid=stelprd3833880

Learn fascinating facts about the Pacific lamprey and their habitat.

Salmon

kids.britannica.com/kids/article/salmon/353740

Explore more facts about a salmon's life with Britannica Kids.

Seahorse Facts

www.theseahorsetrust.org/seahorse-facts/

Discover the amazing abilities of sea horses at this website.

Publisher's note to educators and parents: Our editors have carefully reviewed these websites to ensure that they are suitable for students. Many websites change frequently, however, and we cannot guarantee that a site's future contents will continue to meet our high standards of quality and educational value. Be advised that students should be closely supervised whenever they access the Internet.

INDEX

S

T

V

ABOUT THE AUTHOR

Judy Thorpe studied anthropology at Imperial College London. Through her courses, she developed a keen interest in exploring the ways various cultures' myths and politics impact each other. She is also an animal lover and enjoys learning as much about various animals as possible. When she is not writing, she enjoys rock climbing or curling up with a book and her cat, Punch.